MW01439444

1 MONTH OF FREE READING

at

www.ForgottenBooks.com

By purchasing this book you are eligible for one month membership to ForgottenBooks.com, giving you unlimited access to our entire collection of over 1,000,000 titles via our web site and mobile apps.

To claim your free month visit: www.forgottenbooks.com/free863468

* Offer is valid for 45 days from date of purchase. Terms and conditions apply.

ISBN 978-0-266-54214-8
PIBN 10863468

This book is a reproduction of an important historical work. Forgotten Books uses state-of-the-art technology to digitally reconstruct the work, preserving the original format whilst repairing imperfections present in the aged copy. In rare cases, an imperfection in the original, such as a blemish or missing page, may be replicated in our edition. We do, however, repair the vast majority of imperfections successfully; any imperfections that remain are intentionally left to preserve the state of such historical works.

Forgotten Books is a registered trademark of FB &c Ltd.
Copyright © 2018 FB &c Ltd.
FB &c Ltd, Dalton House, 60 Windsor Avenue, London, SW19 2RR.
Company number 08720141. Registered in England and Wales.

For support please visit www.forgottenbooks.com

"MANU ET MENTE."

A TEXT BOOK

OF

WORKING DRAWINGS OF MODELS

IN

SLOYD

ADAPTED TO AMERICAN SCHOOLS.

BY

GUSTAF LARSSON,

PRINCIPAL OF SLOYD TRAINING SCHOOL, APPLETON STREET, BOSTON, MASS.

PUBLISHED BY THE SLOYD TRAINING SCHOOL.
1893.

TT 188
.L33

PRELIMINARY SLOYD

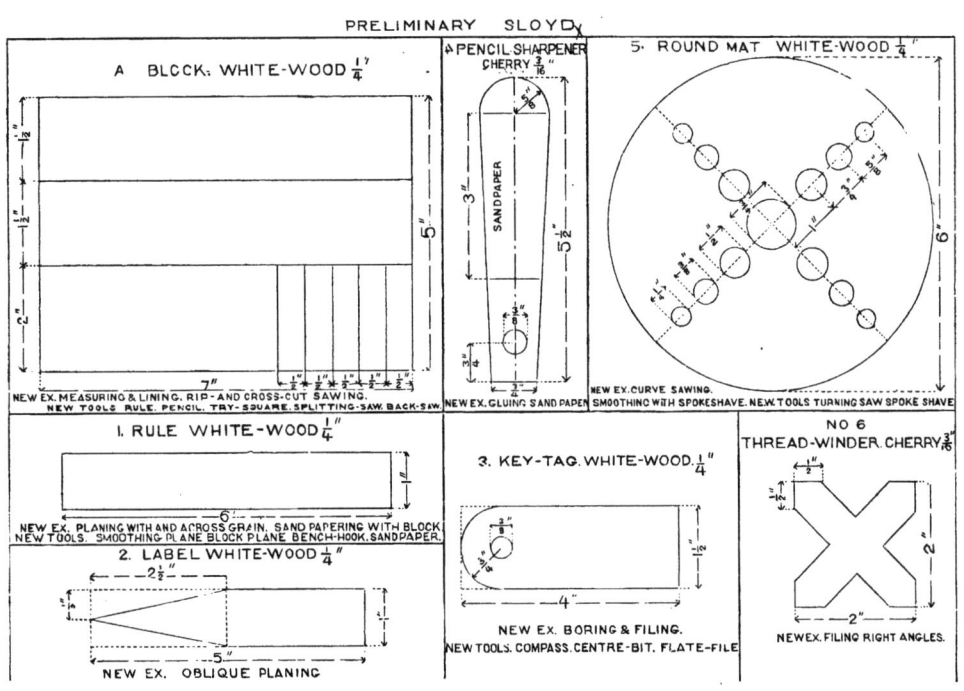

A. BLOCK. WHITE-WOOD ¼"
NEW EX. MEASURING & LINING. RIP- AND CROSS-CUT SAWING.
NEW TOOLS. RULE. PENCIL. TRY-SQUARE. SPLITTING-SAW. BACK-SAW.

4. PENCIL SHARPENER CHERRY 3/16"
NEW EX. GLUING SANDPAPER.

5. ROUND MAT. WHITE-WOOD ¼"
NEW EX. CURVE SAWING. SMOOTHING WITH SPOKESHAVE. NEW TOOLS. TURNING SAW. SPOKE SHAVE.

1. RULE WHITE-WOOD ¼"
NEW EX. PLANING WITH AND ACROSS GRAIN. SAND PAPERING WITH BLOCK.
NEW TOOLS. SMOOTHING PLANE. BLOCK PLANE. BENCH-HOOK. SANDPAPER.

2. LABEL WHITE-WOOD ¼"
NEW EX. OBLIQUE PLANING.

3. KEY-TAG. WHITE-WOOD. ¼"
NEW EX. BORING & FILING. NEW TOOLS. COMPASS. CENTRE-BIT. FLATE-FILE.

No 6 THREAD-WINDER. CHERRY 3/16"
NEW EX. FILING RIGHT ANGLES.

Copyrighted, 1893, by Gustaf Larsson.

SLOYD TRAINING CLASS.

INTRODUCTION.

"Sloid, sloyd (sloid), n. [Sw. *slöjd*, skill, dexterity, esp. mechanical skill, manufacture, wood-carving, = E. *sleight*: see *sleight*².] A system of manual training which originated in Sweden. It is not confined to wood-working, as is frequently supposed (though this is the branch most commonly taught), but is work with the hands and with simple tools. The system is adapted to the needs of different grades of the elementary schools, and is designed to develop the pupils mentally and physically. Its aim is, therefore, not special technical training, but general development and the laying of a foundation for future industrial growth." — *Century Dictionary*.

THIS book presents a plan of the mechanical drawing of the Sloyd models, which are based upon educational principles, briefly stated as follows: —

1. The progression of the exercises should be such as to secure constant and proportionate development of mind and body.

2. Exercises should be so arranged that each model will prepare the way for the next, both physically and mentally.

An "exercise" in Sloyd is a specific use of a tool, involving a certain mental effort.

3. Exercises should always result in a finished article of use. Sloyd claims that by means of a finished model a continuous intelligent interest is obtained, concerning the educational value of which Rev. James Freeman Clarke says: "Real discipline comes to the mind when it acts, not languidly but with full energy, and it acts with energy only when it is interested in what it does." The inspiration which comes from the use of the creative instinct is as valuable in drawing as in any other lesson, and therefore the drawing models should be, as far as possible, objects of real use.

4. The proportions and outlines of the models should be such as to educate the æsthetic sense, and the construction simple enough for the child to reproduce in drawing.

5. The models should admit of a judicious variety of exercises and form.

6. The work should be of such a character as to admit of the best hygienic conditions. The positions assumed in tool-work should counteract, as far as possible, the ill effects of long hours of sitting in school.

The work should be regarded as a regular branch of the school curriculum, and at least two hours weekly should be given to it.

The course of work here described is arranged for four years. The first year's work, which is called "Preliminary Sloyd," is intended for children from eleven to twelve years of age, but may be adapted to those of nine years.

The working-drawings, in the preliminary course (which, as well as in later courses, should always precede the tool-work), are made with one view only (top view or horizontal projection). The models

INTRODUCTION.

represent simple elementary forms, and a few applied geometrical problems. The tool-work for this first year refers only to two dimensions, as the wood is already prepared in thickness. The exercises are represented in fifteen models.

Both tool-work and drawing increase in difficulty as the work progresses. The second year's course gives thirteen models, the third year eleven, and the fourth year seven models.

It has been said elsewhere, and should be remembered, that models of themselves do not constitute sloyd, as paper and clay do not make the kindergarten. It is therefore the prime duty of the teacher to guide the thought of the child, the model being always regarded as the means by which this end is obtained.

The sequence of exercises in the tool-work has been considered and revised after actual experience with classes of children for several years.

These drawings are not accompanied by a full explanatory text, as this book is intended to be used by those teachers who have not only a knowledge of the method of working out each model in wood, but who have a comprehension of the underlying psychological principles on which it is based.

Although a careful supervision has been given to the engraving of the drawings, it has been impossible to avoid certain mechanical defects.

A progressive course, in mechanical drawing apart from tool-work, giving illustrations of objects, with brief directions, is given in an appendix.

I hope that this book may answer satisfactorily some of the needs and questions of my pupils, as well as those of other teachers and friends.

For suggestions and criticism in regard to this course of drawing I am indebted to Prof. Faunce of the Mass. Inst. of Technology.

GUSTAF LARSSON.

Boston, May, 1893.

PRELIMINARY SLOYD.

PUPILS FROM ELEVEN TO TWELVE YEARS OF AGE.

THIS COURSE MAY BE TAKEN BY PUPILS NINE YEARS OF AGE.

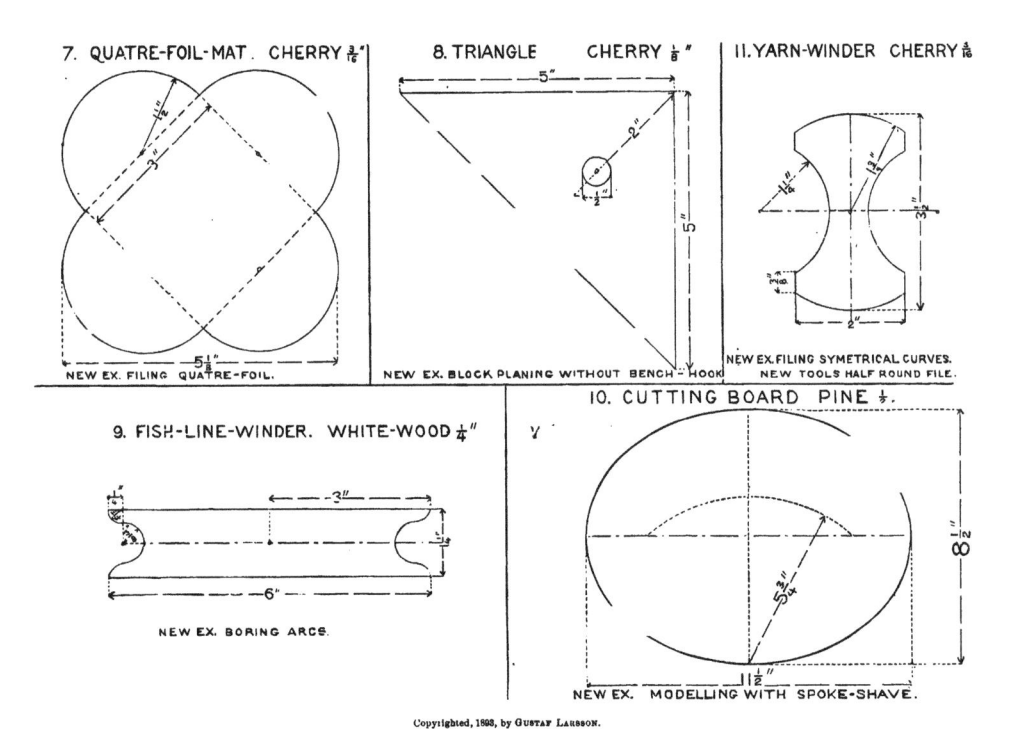

7. QUATRE-FOIL-MAT. CHERRY $\frac{3}{16}"$
NEW EX. FILING QUATRE-FOIL.

8. TRIANGLE CHERRY $\frac{1}{8}"$
NEW EX. BLOCK PLANING WITHOUT BENCH-HOOK

11. YARN-WINDER CHERRY $\frac{3}{16}"$
NEW EX. FILING SYMETRICAL CURVES.
NEW TOOLS HALF ROUND FILE.

9. FISH-LINE-WINDER. WHITE-WOOD $\frac{1}{4}"$
NEW EX. BORING ARCS.

10. CUTTING BOARD PINE $\frac{1}{2}$.
NEW EX. MODELLING WITH SPOKE-SHAVE.

Copyrighted, 1893, by Gustaf Larsson.

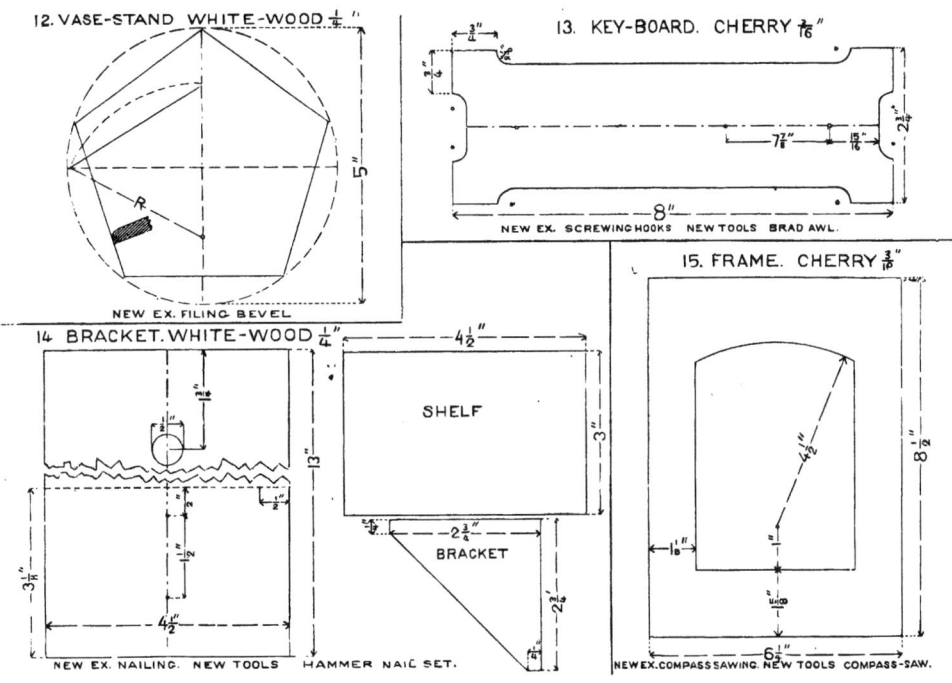

SLOYD.

PUPILS FROM TWELVE TO FIFTEEN YEARS OF AGE.

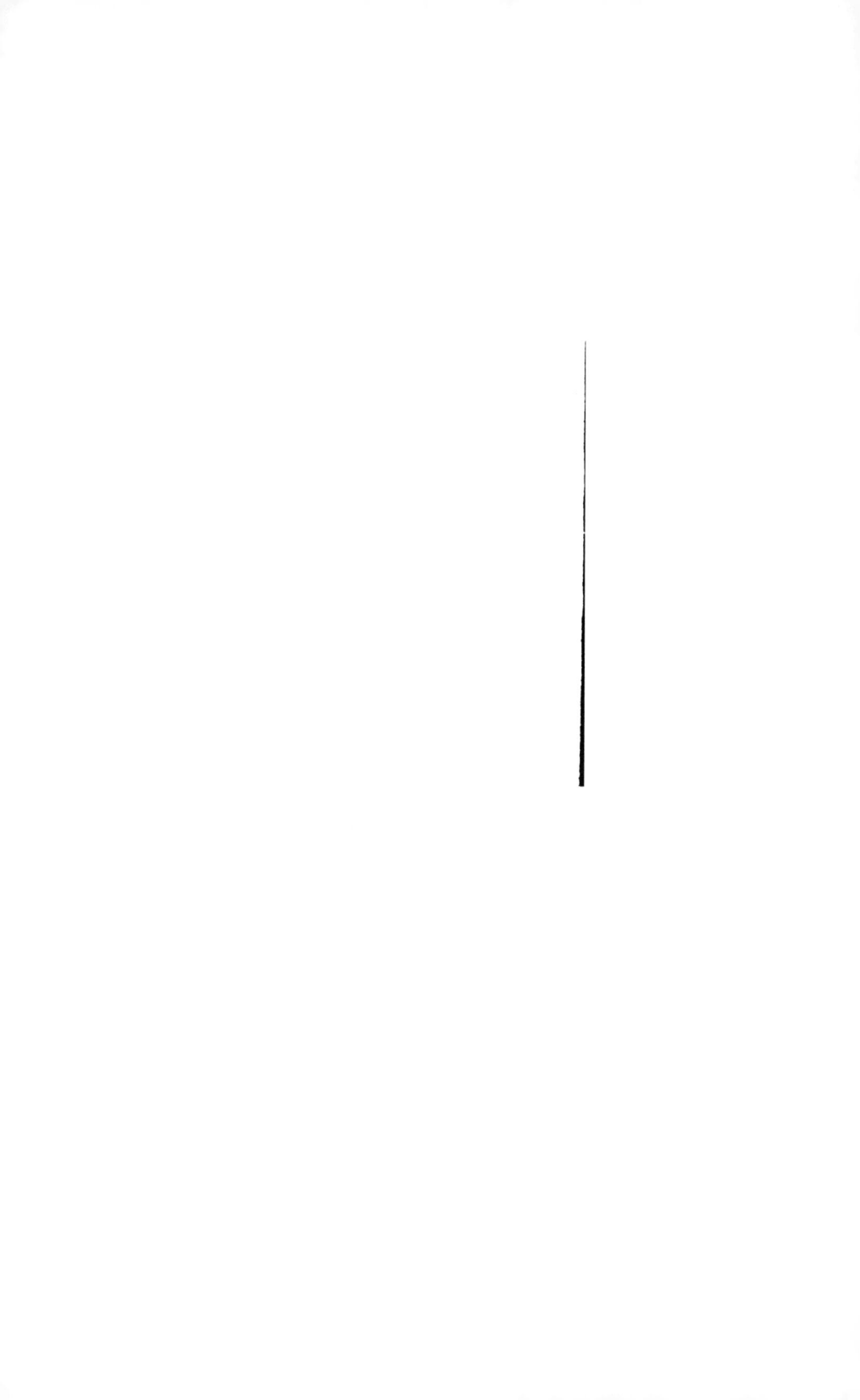

BLOCK

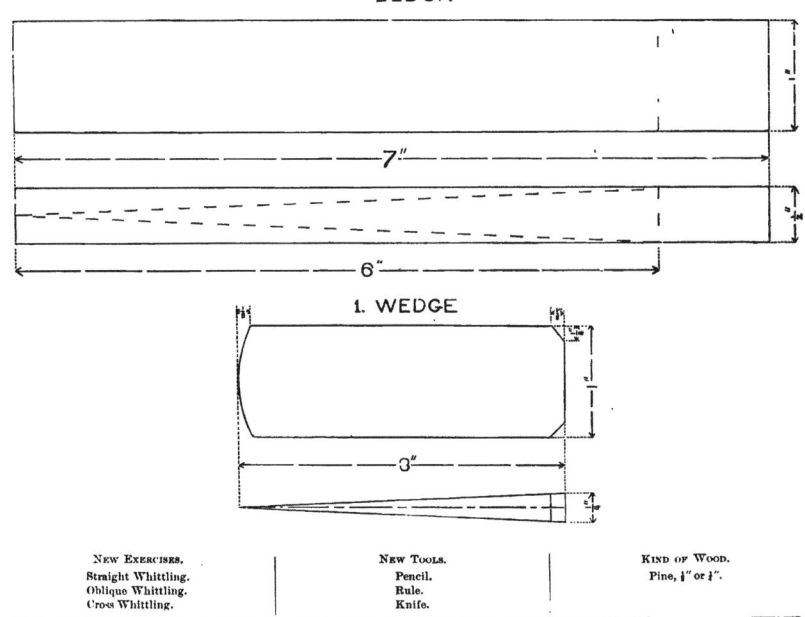

1. WEDGE

New Exercises.	New Tools.	Kind of Wood.
Straight Whittling.	Pencil.	Pine, ½" or ¼".
Oblique Whittling.	Rule.	
Cross Whittling.	Knife.	

Copyrighted, 1893, by GUSTAF LARSSON.

2. FLOWER PIN

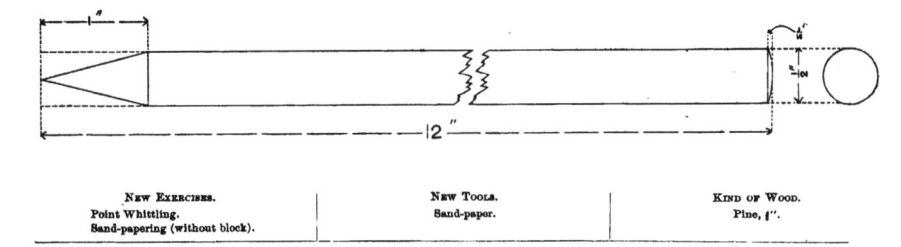

New Exercises.	New Tools.	Kind of Wood.
Point Whittling. Sand-papering (without block).	Sand-paper.	Pine, ⅞".

3. FLOWER STICK

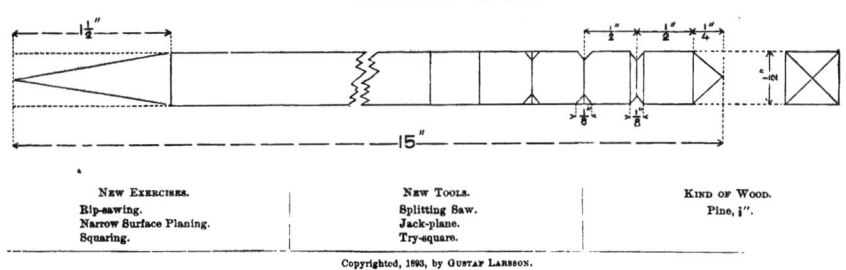

New Exercises.	New Tools.	Kind of Wood.
Rip-sawing. Narrow Surface Planing. Squaring.	Splitting Saw. Jack-plane. Try-square.	Pine, ⅞".

Copyrighted, 1893, by Gustaf Larsson.

4. PENHOLDER

New Exercises.	New Tools.	Kind of Wood.
Boring with Drill Bit. Fitting a Peg. Curve Whittling.	Drill Bit.	Pine, ⅜".

5. TOOL RACK

New Exercises.	New Tools.	Kind of Wood.
Cross-cut Sawing. Gauging. End Planing (in bench-hook). Boring with Auger Bit (vertical). Sand-papering (with block).	Back Saw. Marking Gauge. Block Plane. Bench-hook. Auger Bit.	Pine, ⅞".

Copyrighted, 1893, by Gustaf Larsson.

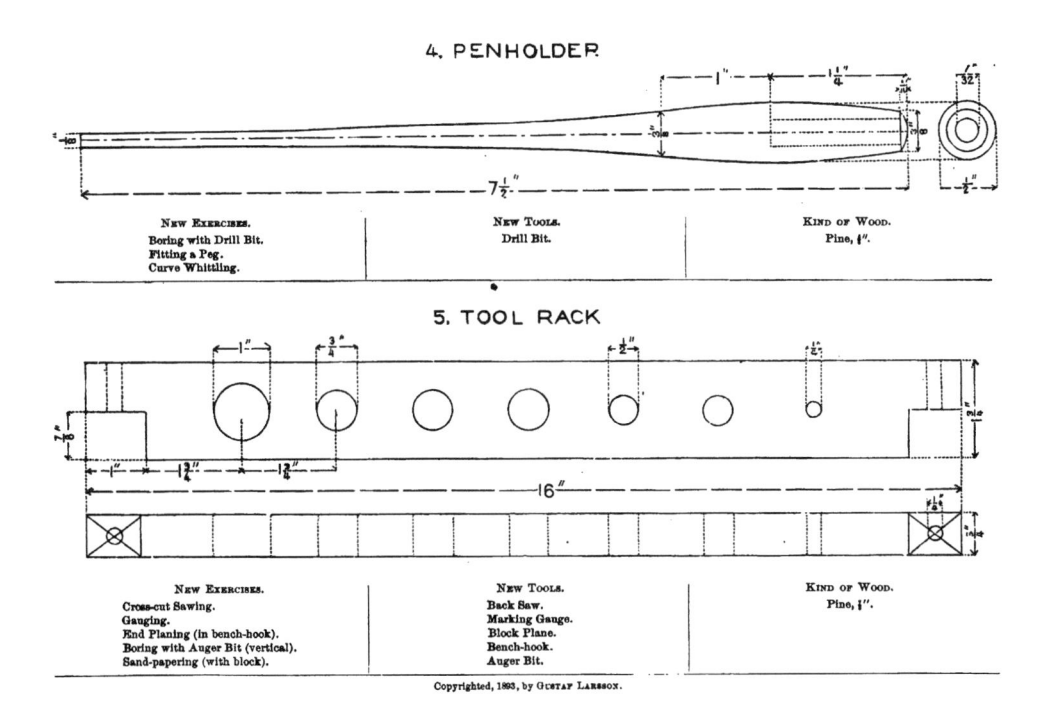

6. COAT HANGER

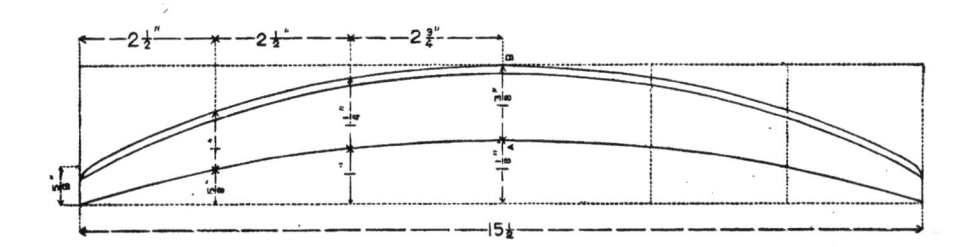

SECTION ON A.B.

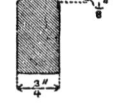

New Exercises.	New Tools.	Kind of Wood.
Curve Sawing.	Turning Saw.	Pine, ⅝".
Smoothing with Spokeshave.	Spokeshave.	
Boring with Brad-awl.	Brad-awl.	

Copyrighted, 1898, by Gustaf Larsson.

7. CUTTING-BOARD

New Exercises.	New Tools.	Kind of Wood.
Broad Surface Planing.	Cutting-off Saw.	Pine, ⅞".
Vertical Chiseling.	Winding Sticks.	
Horizontal Boring.	Chisel.	
Filing.	Flat File.	
End Planing (without bench-hook).		

Copyrighted, 1893, by GUSTAF LARSSON.

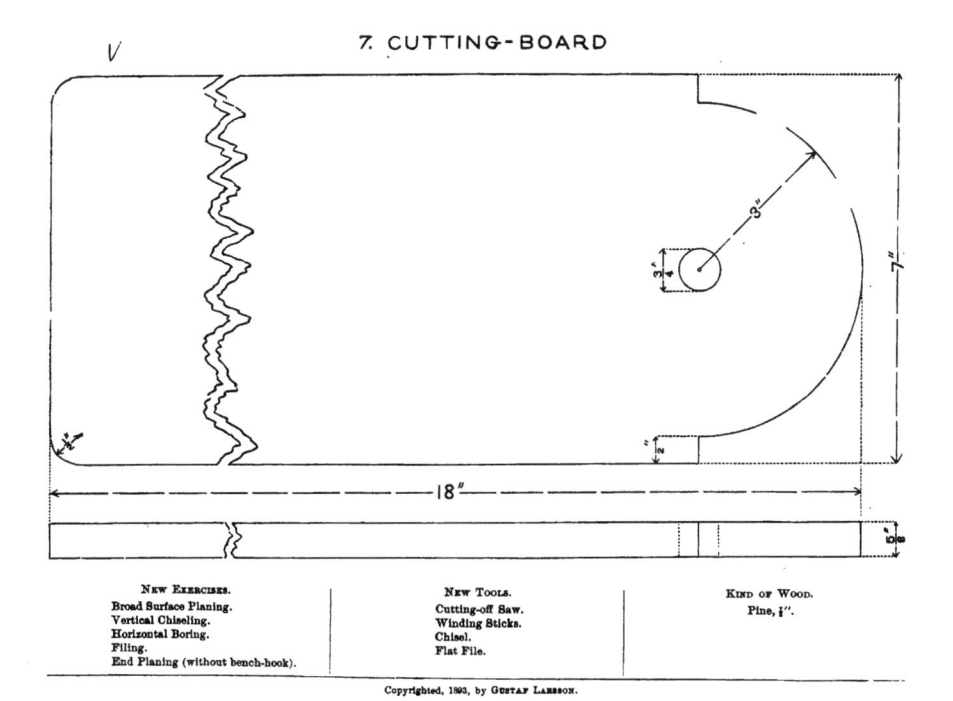

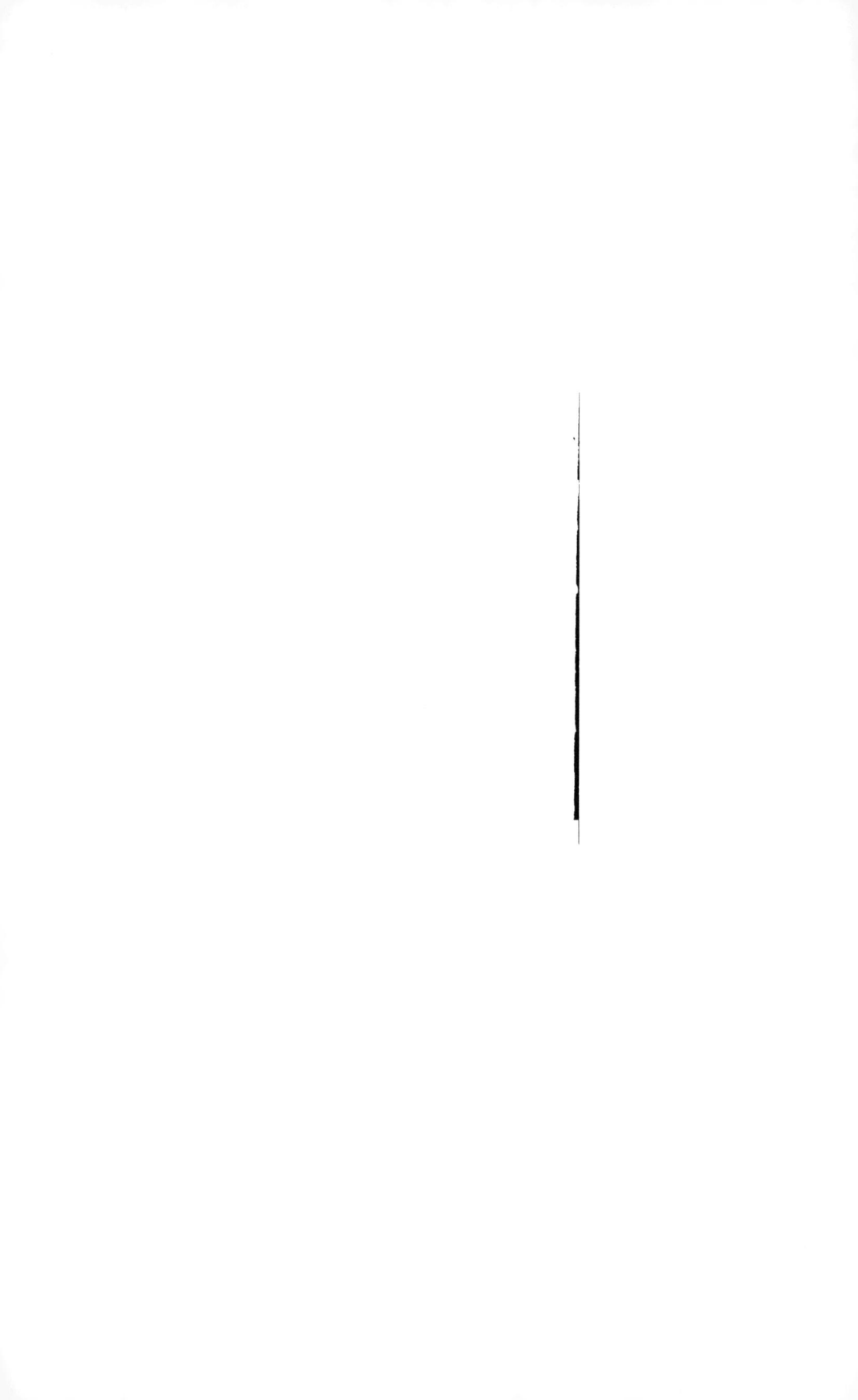

8. FLOWER-POT STAND

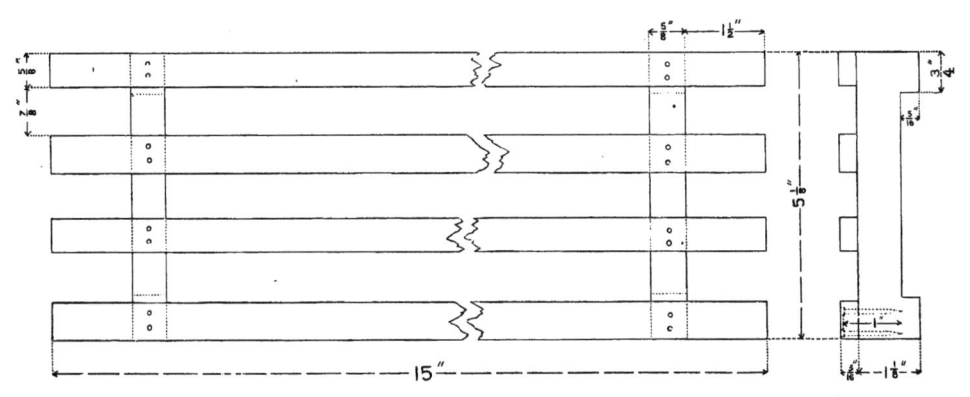

New Exercises.	New Tools.	Kind of Wood.
Nailing.	Hammer.	Pine, ⅜".
Sinking Nails.	Nail Set.	

Copyrighted, 1893, by Gustaf Larsson.

9. FLOWER-POT STOOL

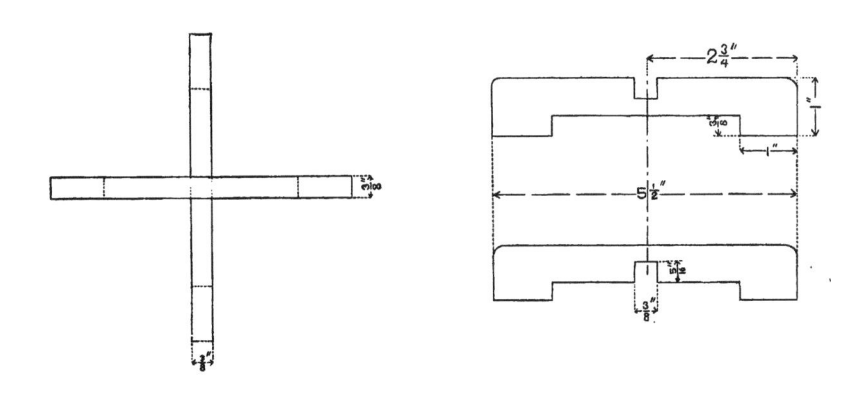

New Exercises.	New Tools.	Kind of Wood.
Making Halved-together Joint.		Pine, ½".

Copyrighted, 1893, by Gustaf Larsson.

10 BENCH-HOOK

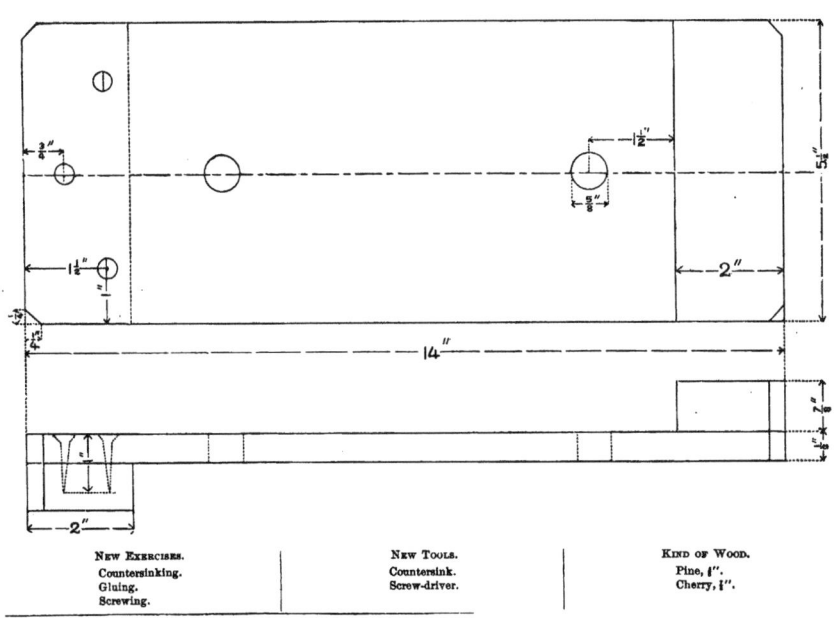

New Exercises.	New Tools.	Kind of Wood.
Countersinking.	Countersink.	Pine, ¾".
Gluing.	Screw-driver.	Cherry, ¼".
Screwing.		

Copyrighted, 1893, by Gustaf Lars

II. HATCHET HANDLE

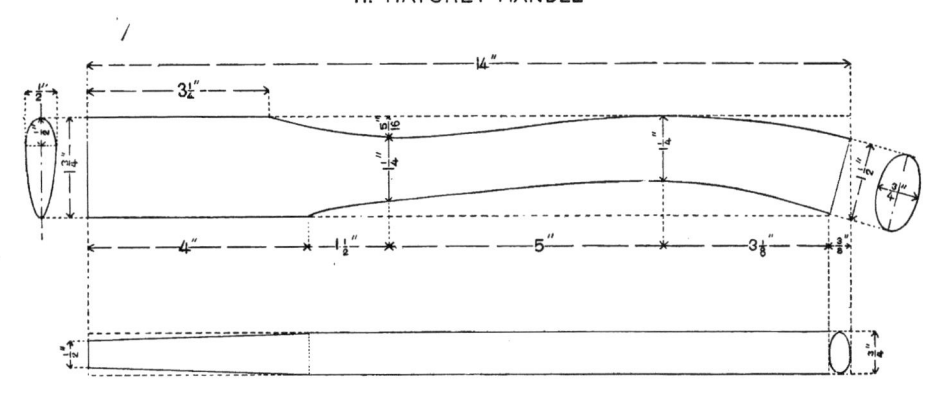

NEW EXERCISES.	NEW TOOLS.	KIND OF WOOD..
Modeling with Spokeshave.	Smoothing Plane.	Beech, 1" rough.
Scraping.	Half-round File.	
	Cabinet Scraper.	

Copyrighted, 1893, by Gustaf Larsson.

12. CORNER BRACKET

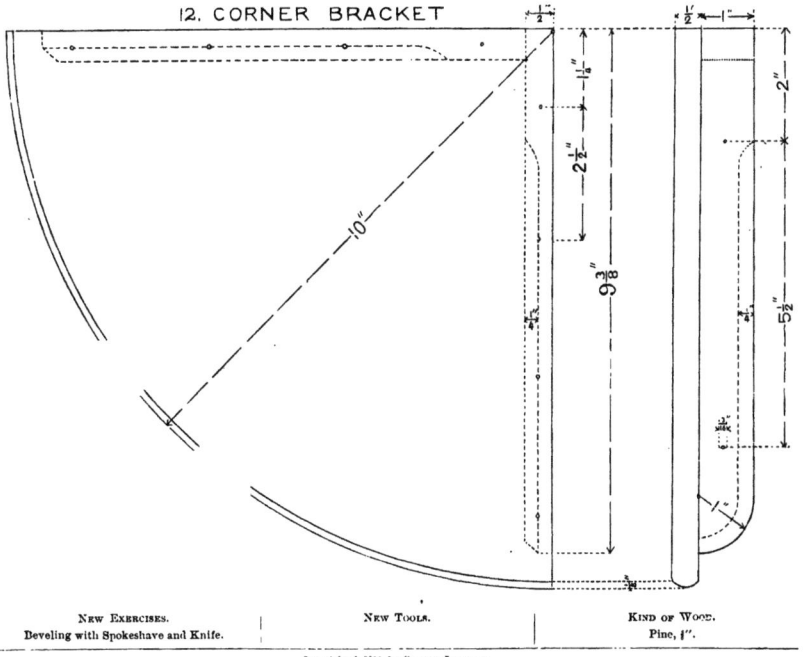

New Exercises.	New Tools.	Kind of Wood.
Beveling with Spokeshave and Knife.		Pine, ⅞".

Copyrighted, 1893, by GUSTAF LARSSON.

13. HAMMER HANDLE

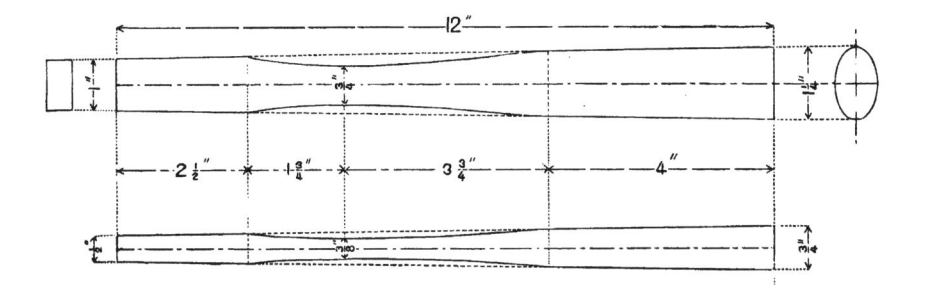

New Exercises.	New Tools.	Kind of Wood.
Oblique Planing.		Beech, 1" rough.

Copyrighted, 1893, by Gustaf Larsson.

14. KEY BOARD

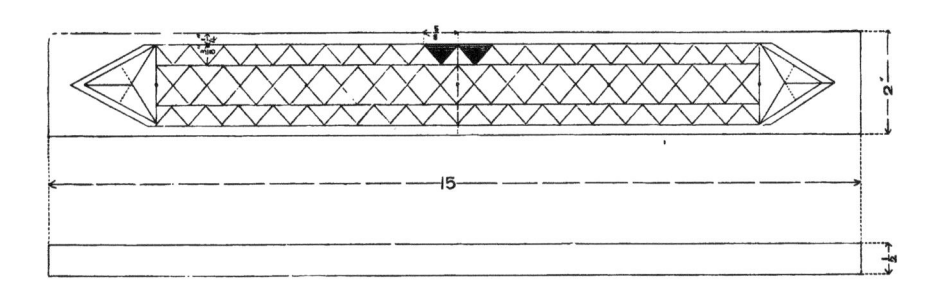

New Exercises.	New Tools.	Kind of Wood.
Spacing with Compass.	Bevel.	Pine, $\frac{1}{2}''$.
Veining.	Veining Tool.	
Carving.	Skew Chisel.	

Copyrighted, 1893, by GUSTAF LARSSON.

15. PAPER KNIFE

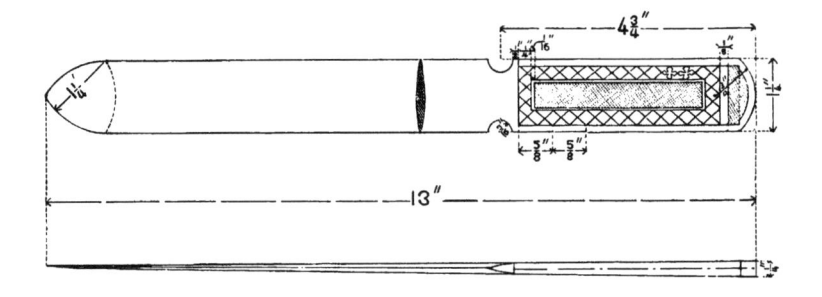

New Exercises.	New Tools.	Kind of Wood.
Wedge Planing.	Round File.	Maple; $\frac{1}{4}$".
Filing Edge.	Carver's Punch.	
Notching.		
Punching.		

d, 1893, by Gustaf Larsson.

16. RULER

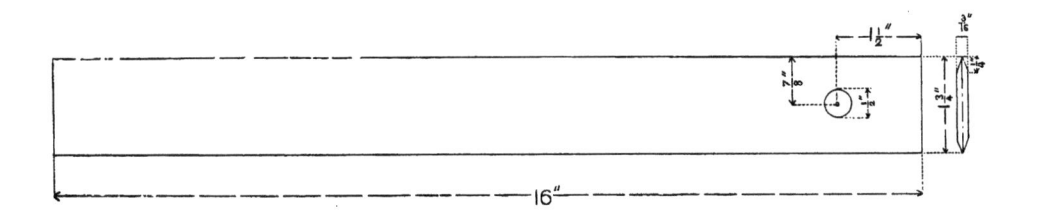

New Exercises.	New Tools.	Kind of Wood.
Beveling Edge with Jack-plane and File. Boring with Centre Bit.	Centre Bit.	Maple, $\frac{1}{4}''$.

Copyrighted, 1893, by Gustaf Larsson.

17. TOWEL ROLLER.

NEW EXERCISES.
Planing a Cylinder.
Fitting Axle.

NEW TOOLS.

KIND OF WOOD.
Pine, 1¼" and ⅞".

IB. FRAME

New Exercises.	New Tools.	Kind of Wood.
Open Mortise and Tenon Joint.	Mortise Gauge.	Pine, 1".
Making and Fitting Dowels.	Mallet.	

Copyrighted, 1893, by Gustaf Larsson.

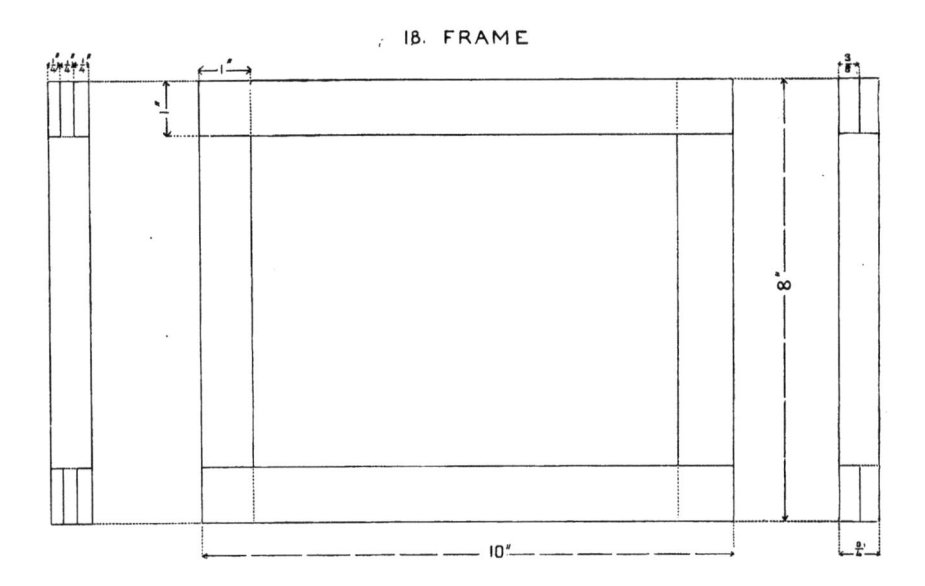

19. BOX

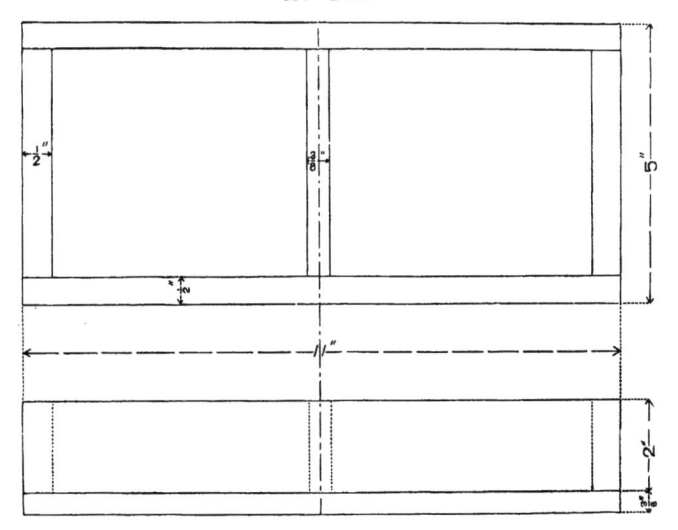

NEW EXERCISES.	NEW TOOLS.	KIND OF WOOD.
Fitting and Nailing Square Joints.		White Wood, 1".

Copyrighted, 1893, by GUSTAF LARSSON.

20. PEN TRAY

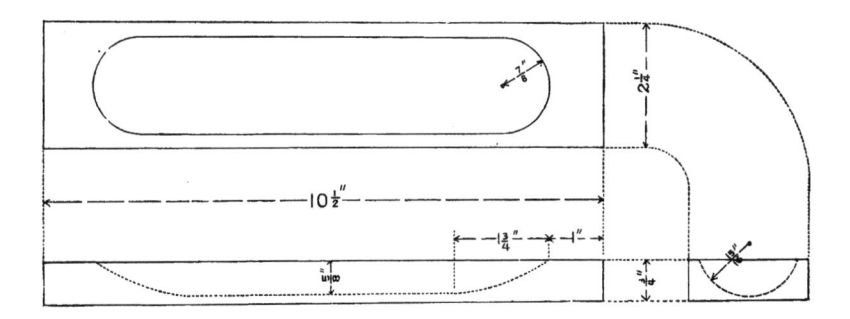

New Exercises.	New Tools.	Kind of Wood.
Grooving with Gouge.	Gouge. Round Cabinet Scraper.	Gum, 1″.

21. HAT RACK

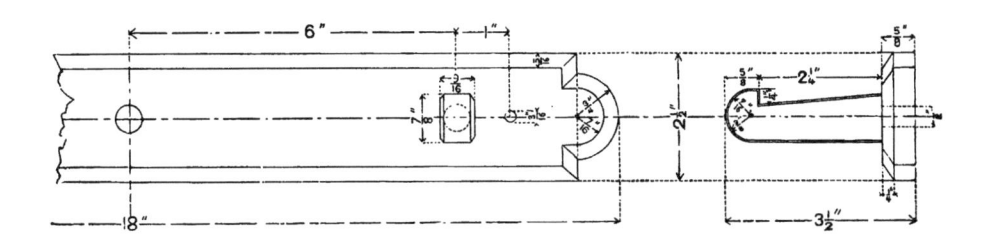

NEW EXERCISES.	NEW TOOLS.	KIND OF WOOD.
Chamfering. Straight-edge Beveling.		Pine, ⅞".

22. PICTURE FRAME

New Exercises.	New Tools.	Kind of Wood.
Half Lapping. Grooving with Chisel.		Gum, ¼".

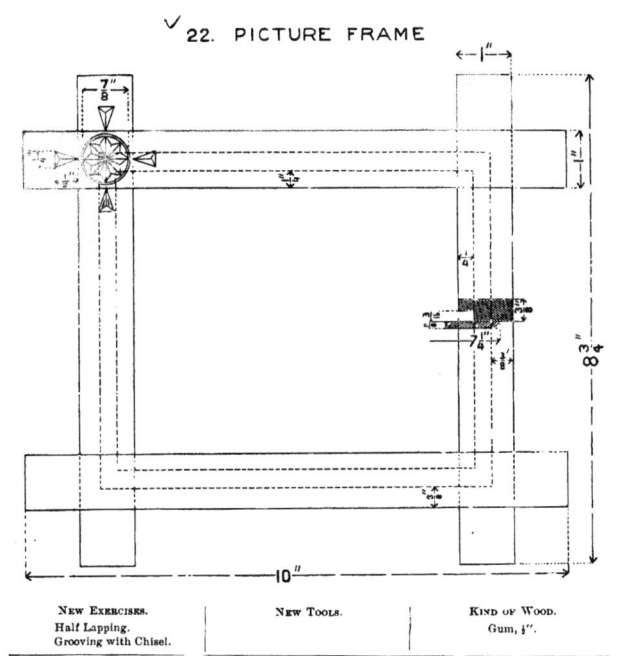

Copyrighted, 1893, by Gustaf Larsson.

23. CAKE SPOON

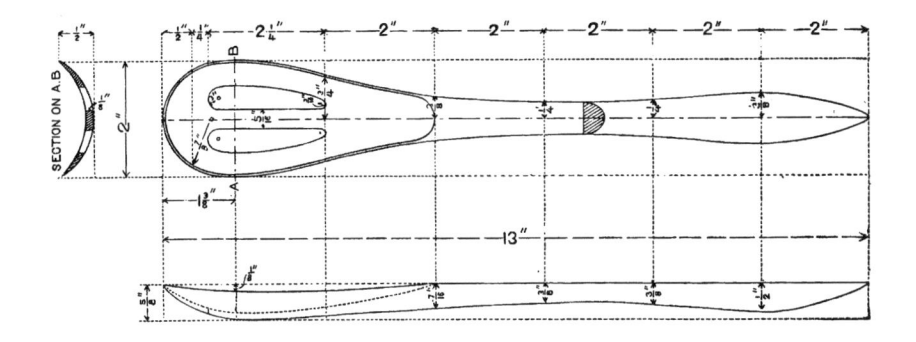

New Exercises.	New Tools.	Kind of Wood.
Compass Sawing.	Compass Saw.	Cherry, ¾".

Copyrighted, 1893, by Gustaf Larsson.

24. PICTURE FRAME

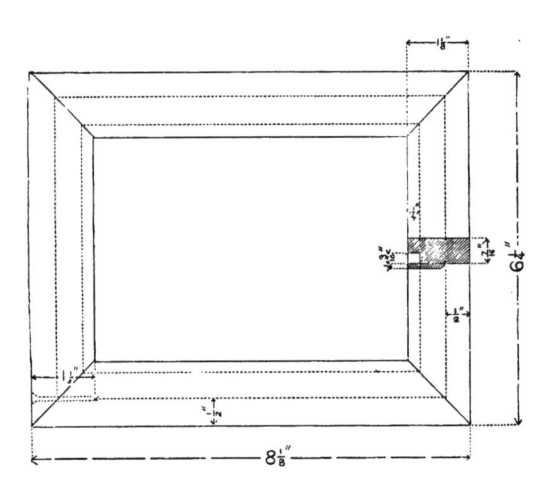

New Exercises.	New Tools.	Kind of Wood.
Grooving with Rabbet-plane. Mitring.	Rabbet-plane.	Cherry, ¼".

Copyrighted, 1893, by Gustaf Larsson.

25. FOOT STOOL

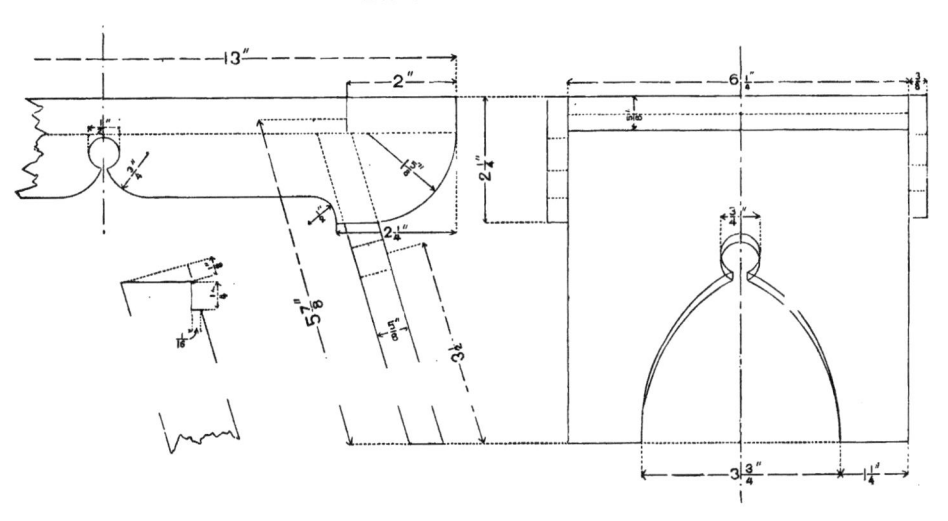

New Exercises.	New Tools.	Kind of Wood.
Half-oblique Dovetail.		Pine, ⅞" and 1¼".

Copyrighted, 1893, by Gustaf Larsson.

26. SCOOP

New Exercises.	New Tools.	Kind of Wood.
Vertical Gouging.	Gouge, bevel inside.	Cherry, 2″.
Cutting with Drawing Knife.	Drawing Knife.	

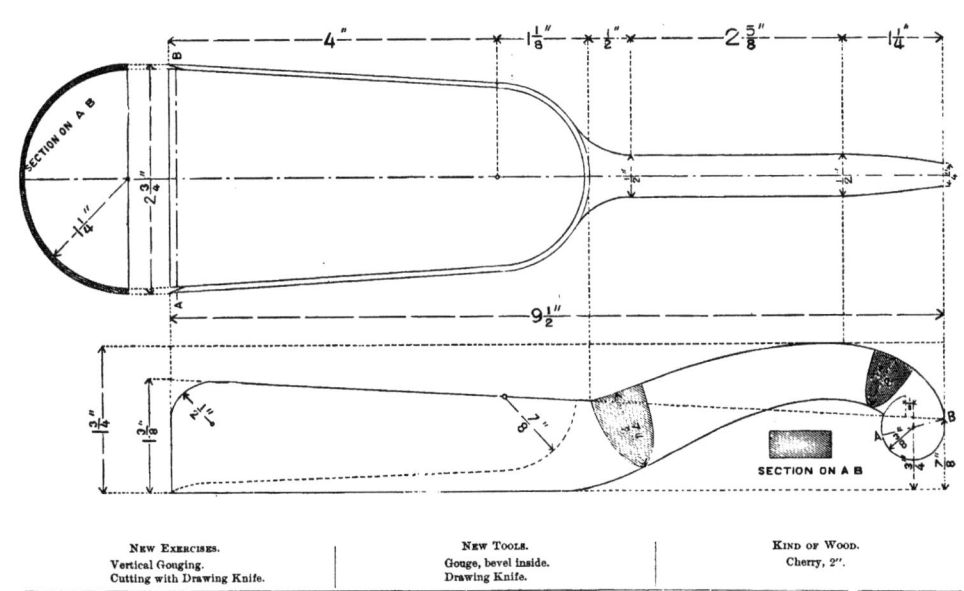

27 A. BRACKET

27 B. BOOK RACK

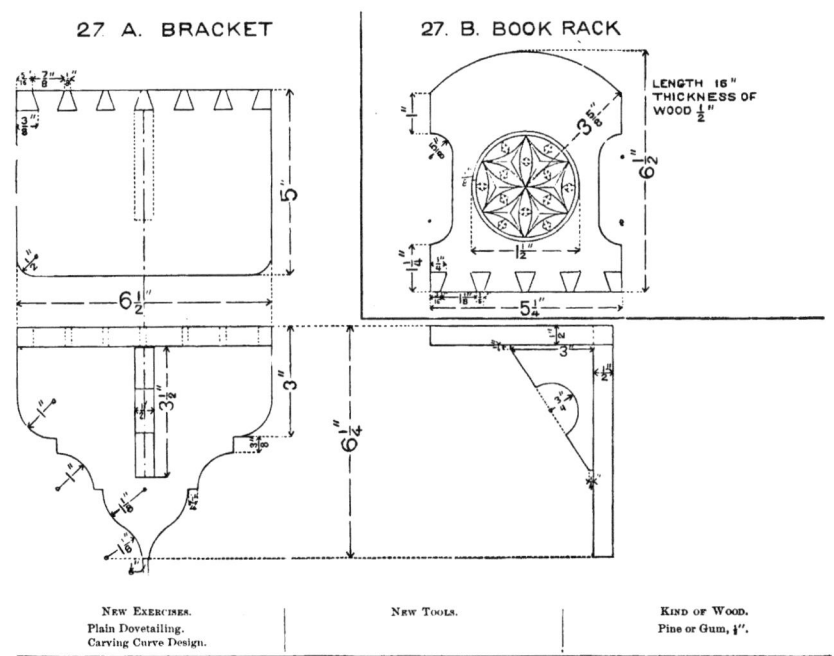

LENGTH 16"
THICKNESS OF WOOD ½"

NEW EXERCISES.
Plain Dovetailing.
Carving Curve Design.

NEW TOOLS.

KIND OF WOOD.
Pine or Gum, ½".

Copyrighted, 1893, by GUSTAF LARSSON.

28. KNIFE BOX

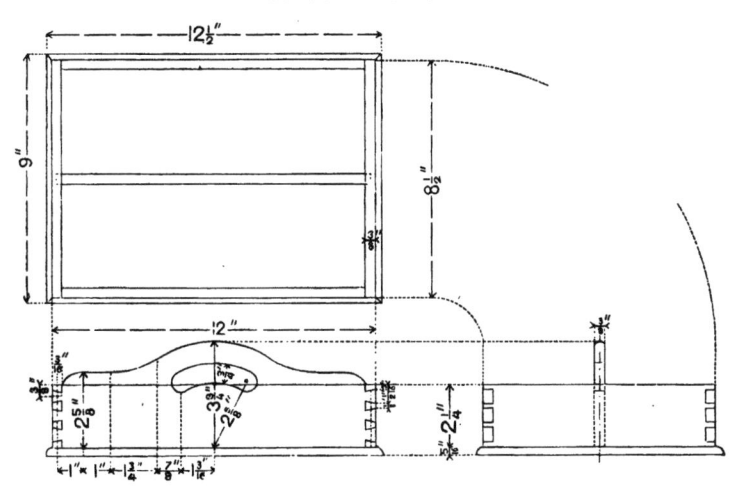

New Exercises.	New Tools.	Kind of Wood.
Plain Jointing.	Jointer Plane.	Pine, ¼".
Square Grooving.		
Quarter-round Beveling with Plane.		

Copyrighted, 1893, by GUSTAF LARSSON.

29   DRAWING BOARD

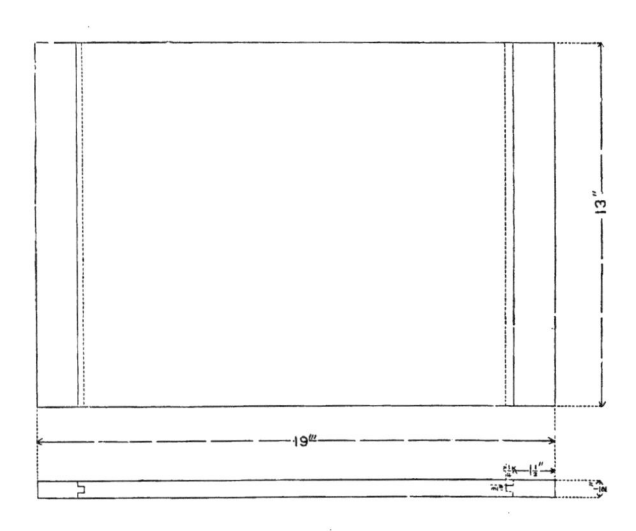

New Exercises.	New Tools.	Kind of Wood.
Matching.	Matching Plane.	Pine, ⅞".
Cleating.	Cabinet-maker's Clamps.	

Copyrighted, 1893, by GUSTAF LARSSON.

30 TRAY

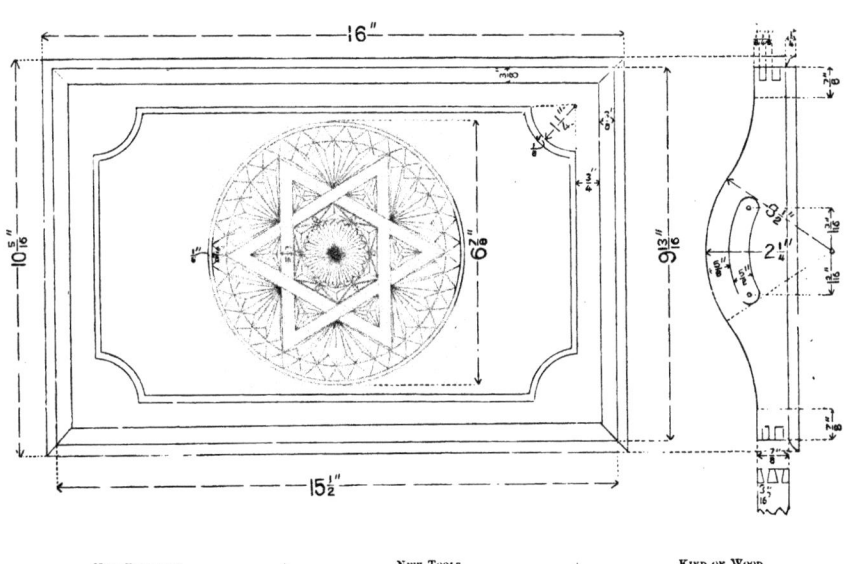

New Exercises.	New Tools.	Kind of Wood.
Dovetailing with a Mitre. Shellacing.	Parting Tool.	Pine, ⅞".

Copyrighted, 1893, by GUSTAF LARSSON.

31. TOOL-CHEST

NEW EXERCISES.	NEW TOOLS.	KIND OF WOOD.
Panel Grooving.	Mitre Box.	Pine, $1''$, $\frac{1}{2}''$, and $\frac{1}{4}''$.
Half Blind Dovetailing.	Mitre Plane.	
Blind Mortise and Tenon Joint.	Framing Chisel.	
Fitting Hinges and Lock.	Plow.	

Copyrighted, 1893, by GUSTAF LARSSON.

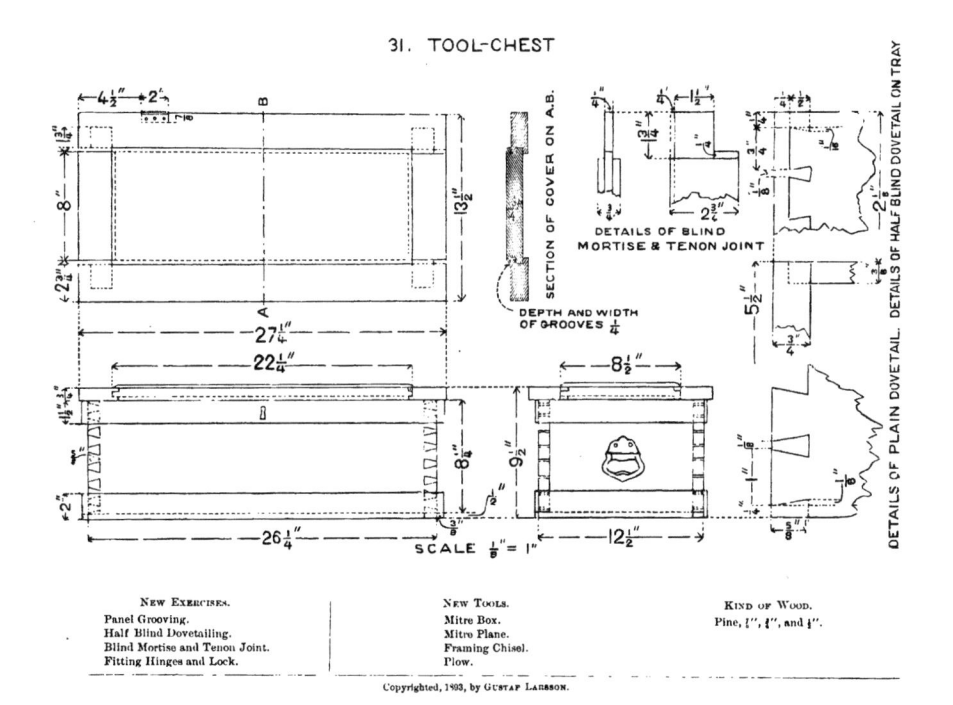

WORKING POSITIONS WITH A FEW GENERAL DIRECTIONS.

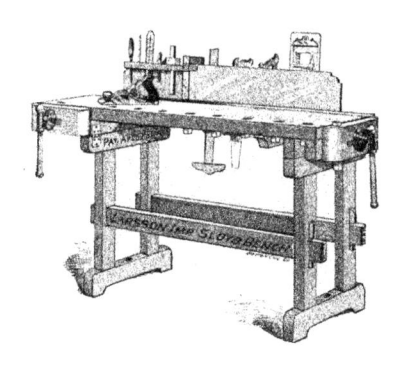

SLOYD BENCH.

WHITTLING.

1. Grasp the knife in the right hand, with the thumb bound over the fingers, as in clenched fist.
2. Hold the wood at the end nearest you.
3. Rest the forearms against the body, and cut from you and downward.
4. Do not cut from the very end, but start just beyond the hand, and turn the wood to finish.
5. Try to use the whole length of the blade by drawing it through the wood as you cut. Do not scrape.
6. If you have a broad face to cut, take off the edges first.
7. Never cut clear across an end, but always from the sides toward the middle.

SAWING.

1. The saw should be grasped firmly, with one hand round the handle and the other hand resting on the wood.
2. In starting a cut, the saw should be drawn across the wood toward the body two or three times, the thumb serving as a guide for the saw.
3. The saw must not be pressed down upon the work, but moved horizontally, with a long, light, and even stroke.
4. The turning-saw should be grasped with both hands, as near the handle as possible.

PLANING.

Grasp the plane by the handle and press and guide firmly with the other hand upon the " toe," taking special care to keep the horizontal position of the plane at the ends of the board.

In planing a piece true, these six rules should be followed: —

1. Plane one broad face smooth and true.
2. Finish one narrow face at right angles to the first.
3. Take width, and plane.
4. Take thickness, and plane.
5. Blockplane one end true.
6. Take length, and blockplane.

BORING.

1. Fasten the bit firmly in the bit brace.

2. Be sure the bit is perpendicular to the surface of the wood; it may be tested by the try square.

3. To keep the bit perpendicular in vertical boring, move around slowly, looking down upon the bit while turning the brace.

4. When the centre point appears, stop boring and finish from the opposite side.

VERTICAL CHISELLING.

1. Grasp the handle of the chisel with one hand, hold the wood and guide the chisel with the other.
2. Hold the flat side of the chisel next the wood and make a clean cut.
3. Cut but a little at a time, and notice the direction of the grain.
4. Be sure to avoid a cramped position of the chest.

FILING.

1. Fasten the work in the vice.
2. Grasp the file handle firmly with one hand, the other hand at the end of the file, with only the thumb on top.
3. Move the file nearly its whole length across the work with a steady stroke.
4. Keep the line and test with try square.

APPENDIX.

THIS course in drawing is intended for use in elementary schools (children about 13 years of age) and includes all the important facts necessary in making a simple working drawing. Technical terms have been avoided as far as possible and the ordinary method of teaching projection somewhat simplified.

In applying this course the following points should be carefully considered:—

1. Pupils should be led to see that drawing is a convenient and forcible means of thought expression.

2. In teaching orthographic projection the third angle has been employed; that is, the object is placed below the horizontal plane and behind the vertical plane.

3. A working drawing should contain only such views, lines, and dimensions, as are actually necessary for a clear comprehension of the object to be made.

4. The objects used should afford variety, and also sufficient repetition of principles to impress them upon the mind of the pupil.

5. As a rule no object should contain more than four new facts.

6. All the objects drawn should be made to exact dimensions, and should be as far as possible articles of recognized utility.

Copying from drawings should be avoided, and drawing from dictation *sparingly* practised. It is desirable to have a model for each pupil while drawing the first six models. When these are drawn, a smaller number of objects may be sufficient for a class, if the pupil is required to measure and make a freehand working sketch of the model as a basis for the working drawing.

It is preferable to make the drawings full size. Where a drawing to scale (reduced size) is necessary, the actual dimension should be given.

The following instruments, of good quality, should be furnished each pupil: Drawing-board 19″ × 13″, T-square, Triangle 90° and 45°, One foot-rule, Compasses, Drawing-paper 18″ × 12″, Hard pencil, Thumb-tacks, and Rubber.

The first lesson given should relate to the proper use of the instruments, and should be illustrated, clearly, on the blackboard by the teacher. For exercises to be drawn, see plate A.

GENERAL DIRECTIONS.

1. The drawing-board should be kept with one long side towards the person drawing.

2. In drawing horizontal lines the T-square is used. Vertical lines are drawn with triangle and T-square in connection.

3. The T-square is placed with head close to left hand side of the board and only its upper edge is used.

4. Hold the compasses at the top between thumb and forefinger only, keeping the needle point as nearly perpendicular to the surface as possible.

5. Use a hard pencil which, in drawing straight lines, should be sharpened to a thin edge, somewhat like a wedge.

6. The drawing-paper is fastened with thumb tacks in its upper corners stretched on board true to T-square.

7. The border lines are drawn first, making a rectangle of given size. If the paper is 18″ × 12″ draw this rectangle 16″ × 10″. In drawing the border lines proceed thus: Measure and draw a horizontal line 1″ from the top. Measure on this line 1″ from the left edge and draw a vertical line. From the point where these two lines intersect measure the length of the rectangle on the horizontal line and its width on the vertical line. Complete the rectangle by lines drawn through these points.

8. A neat uniform arrangement of name of pupil, school, and object drawn, written on the drawing is recommended throughout the course.

LINES.

Outline, solid : ─────────────

Dimension lines, one-half inch dash : ─── ─── ───

Extension and construction lines, one-eighth inch dash : — — — —

Lines for invisible edges, one-thirty-second inch dash : - - - - - - - -

Centre and section lines, dot and dash . ─── . ─── . ─── . ───

THE MODELS
For the Course of Drawing have been arranged according to the Following Plan.

I. Drawings with one view (top view or horizontal projection).
  No. 1, ruler; No. 2, label; No. 3, key-tag; No. 4, pencil sharpener.
II. Drawings with two views.
  No. 5, sand-paper block; No. 6, ruler; No. 7, steps; No. 8, moulding.
III. Drawings with two views and incomplete section.
  No. 9, paper-knife; No. 10, coat-hanger.
IV. Drawings with three views.
  No. 11, tool rack; No. 12, pen tray.
V. Drawings with complete sections.
  No. 13, spool.

NOTE.— This course is arranged for the grammar grades, when the children are supposed to have had some previous training, such as paper cutting and folding and clay modelling.
Every teacher will understand that the course here outlined is simply a skeleton of principles, which may be filled out to suit the different capacities of classes.

MODEL No. 1.
(Rule.)

New Facts. Measuring object; drawing oblong; use of dimension and extension lines; dimensioning.

Directions: 1. Top view is a representation of an object as seen when looked down upon in its natural position (seen through the horizontal plane). One view is generally sufficient in a working drawing of an object of uniform given thickness of such material as paper, pasteboard, or thin wood.
2. The drawing should be properly placed on the paper.
3. In drawing this model, notice the directions for drawing the border line.
4. Place dimension lines ($\frac{1}{8}''$ dash) about $\frac{1}{4}''$ outside drawing, leaving space in the middle of the line for the dimension, written with clear numbers reading from left to right and from bottom to top. As a general rule place the dimensions below the top view and at its right hand side.
5. To limit the dimension indicated by the numbers, arrow-heads are placed at each end of the dimension line with points touching extension lines ($\frac{1}{8}''$ dash.)

MODEL No. 2.
(Label.)

New Facts. Dimensioning oblique lines; writing fractions; use of construction lines.

Directions: 1. In order to impress upon pupil's mind the convenience of drawing as a means of thought expression have him describe in words Model 2.
2. Consider the type form nearest related to the object and draw that with light lines.
3. Let the fraction line be in line with the dimension line thus: ——— $1\frac{1}{2}''$ ——— not ——— $1\frac{1}{2}''$ ———.

MODEL No. 3.
(Key-tag.)

New facts. Dimensioning circle and arc, tangent.

Directions: 1. Dimension the diameter of a circle, putting the dimension outside circle between extension lines.
2. Dimension radius of an arc placing a small circle around the centre which will take the place of an arrow-head, the dimension line touching the circumference of that circle.
3. Place the dimension line of an arc at 45°, the dimension reading upwards.
4. Dimension lines should not cross each other.

MODEL No. 4.
(Pencil Sharpener.*)

New Facts. Use of centre-line.

Directions: 1. Centre lines (dot and dash lines) are to be used when the relation of certain facts to the centre must be defined.
2. In drawing objects with symmetrical outlines draw the centre line first.

MODEL No. 5.
(Sand-paper block.)

New Facts. Drawing of two views (two planes), top view and front view; horizontal intersection line; point of tangency.

Directions: 1. Front view is a representation of an object standing on its natural base, looked at from the front.

* A piece of No. 0 sandpaper should be glued on the surface.

APPENDIX.

2. Draw a horizontal line at stated distance from the top border-line. This line is called intersection line.

3. Draw the top view of the model about ½" above that line. Extend its vertical outlines downwards and draw the front view between these lines about ½" below the intersection line.

4. When a curve and flat surface are tangent the intersection should *not* be represented by a line.

NOTE.— To illustrate the reason for the position of the views, fold the paper at right angle at the intersection line with the drawing outside. Then if the paper is placed above the object, the views drawn will correspond to the faces of the object as they would appear if seen through the paper.

MODEL No. 6.
(Ruler.)

New Facts. Necessity of front and side (or end) views. Vertical intersection line. How to represent a break.

Directions: 1. Side view is a representation of an object as seen through the profile plane.

2. Draw a vertical intersection line at a given distance from the border line.

3. Objects too long to be drawn on the regular size paper may be represented as broken when that which is given indicates the character of the omitted portion.

The way to represent this break may be best illustrated by breaking a straight-grained piece of wood.

MODEL No. 7.
(Steps.)

New Facts. Sketching; front and side (or end) views; measuring and dimensioning spaces.

Directions: 1. Decide, by sketching, which views give the clearest representation of the model.

2. Dimension the sketch in following order: First, "over all" dimensions, length, width, and thickness; second, smaller parts.

3. In dimensioning adjoining spaces, let arrow-heads form a cross thus : <——×— - —×——>

4. Put dimensions in places easiest to read; do not repeat dimensions.

NOTE.— Always bear in mind that the facts and dimensions needed to *make* an object are the same as are required in drawing it.

MODEL No. 8.
(Moulding.)

New Facts. Intersection of curve and plane surfaces; convenience of drawing side view first. Quadrant.

Directions: 1. Locate a vertical intersection line.

2. In drawing the side view locate a vertical line representing the width and base of the moulding.

3. Draw a horizontal centre line for the side view.

4. Measure off thickness of moulding on this line.

5. Draw semicircle, spaces, and quadrants as measured from the model.

6. Draw vertical outlines of the front view.

7. Project and draw horizontal lines on the front view from points in the side view where lines intersect.

MODEL No. 9.
(Paper Knife.)

New Facts. Cross section on the figure; locating corners of an equilateral triangle.

Directions: 1. Intersection of curved and flat surfaces not tangent should be indicated by a line

2. In drawing the point of the blade, the teacher will have an opportunity to show the construction of an equilateral triangle.

3. Draw section on blade.

When the form of an object cannot be shown in the common views, a section is taken. A section is an imaginary cut showing interior parts, and is always perpendicular to the vertical or horizontal plane. A section should be cross hatched; that is, covered with parallel oblique lines, usually at 45°. When convenient, this may be put on the drawing where the cut is supposed to be made. When the part cut is all that is represented, the section is called *incomplete*. A *complete* section shows not only the parts cut, but that which is seen beyond.

MODEL No. 10.
(Coat-hanger.)

New Facts. Method of drawing freehand curves; how to represent an incomplete section.

Directions: 1. Measure the height of the model by placing it upon a plane surface (the curved top is measured with rule and T-square in connection).

2. Draw an oblong with light outlines the same length as the model, and width equal to its height.

3. To facilitate the measuring of the model some lines perpendicular to its base may be drawn on half of it at stated distances from the centre.

4. Draw in the oblong lines corresponding to these on each side of centre line.

5. Measure off on these lines the perpendicular distances from the base of the model to the lower curve at the lines drawn on the model.

6. Locate points for the upper curve by measuring the width of the model on the lines.

7. Locate corresponding points in the other half of the oblong.

8. Draw the curve.

9. A top view and an incomplete section taken at the middle of the length are now drawn. To make the drawing clear this section is placed outside the view, a section line indicating where it is taken.

MODEL No. 11.
(Tool rack.)

New Facts. Necessity of three views (three planes); how to represent invisible edges; intersection of horizontal and oblique surfaces.

Directions: 1. Locate and draw horizontal and vertical intersection lines.

2. Draw outlines of top and front views at equal distances from intersection lines.

3. Project outlines of top view to the vertical intersection line.

4. Find corresponding points on the horizontal intersection line by means of arcs with the point where the intersection lines meet as centre.

5. By projecting from these points and the front view draw the side view.

6. Invisible edges are represented by $\tfrac{1}{16}''$ dash lines. (See note.)

7. Illustrate the reason for representing invisible edges.

8. Avoid as much as possible to dimension parts on invisible lines.

NOTE.— In this model the invisible lines in the end view may be left out, unless the pupil discovers and desires to have them in; they are not necessary in a *working* drawing of this model.

MODEL No. 12.
(Pen Tray.)

Review of previous model.

MODEL No. 13.
(Spool.)

New Facts. Exercises in drawing complete sections; measuring a cylindrical object with rule; drawing view partly in elevation and partly in section.

Directions: 1. Draw top view first, and project the front view. To find the smaller diameter of the spool, apply a straight edge to the length of the spool, measure the space between the straight edge and the smaller cylinder. Twice this space, subtracted from the larger diameter, gives the smaller diameter.

2. To measure the bevel of the spool, apply a right angle at the intersection of the smaller cylinder and the bevel. Measure the space between the outer edge of the bevel and the right angle.

Complete section, see Model No. 9.

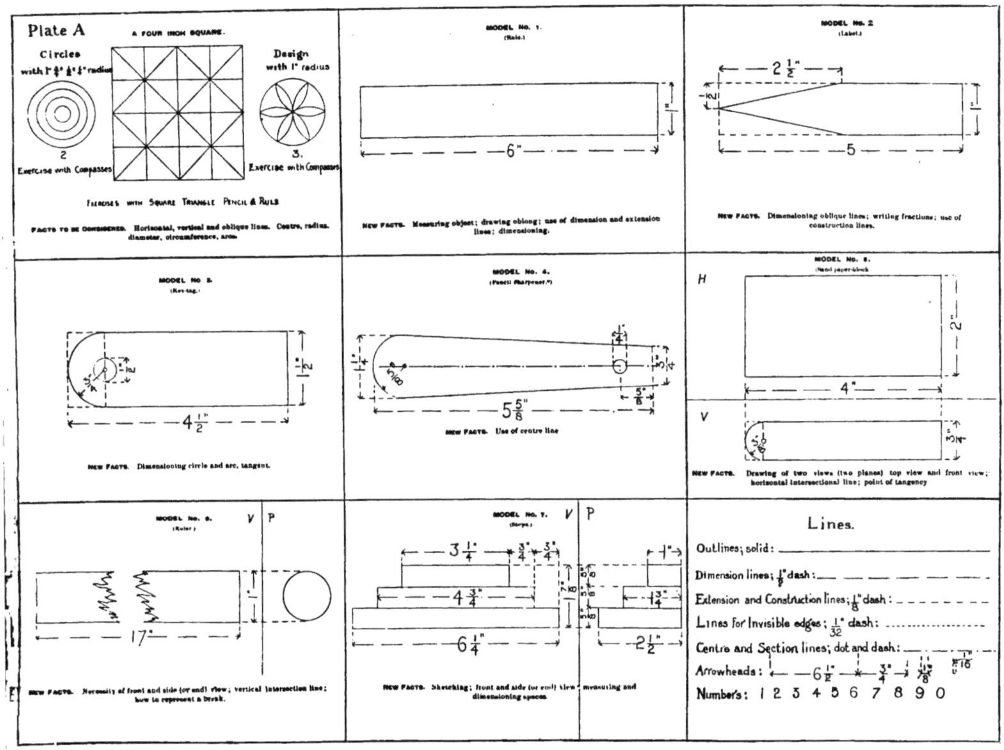

Plate A
A FOUR INCH SQUARE.
Circles
with 1" ¾" ½" ¼" radius
2
Exercise with Compasses
Design
with 1" radius
3.
Exercise with Compasses
Exercises with Square Triangle Pencil & Rule
FACTS TO BE CONSIDERED. Horizontal, vertical and oblique lines. Centre, radius, diameter, circumference, arcs.
MODEL No. 1.
(Rule.)
1"
6"
NEW FACTS. Measuring object; drawing oblong; use of dimension and extension lines; dimensioning.
MODEL No. 2
(Label.)
2 ½"
½"
1"
5
NEW FACTS. Dimensioning oblique lines; writing fractions; use of construction lines.
MODEL No. 3.
(Key-tag.)
½"
¾"
1 ½"
4 ½"
NEW FACTS. Dimensioning circle and arc, tangent.
MODEL No. 4.
(Pencil Sharpener.)
¼"
1 ¼"
⅝
¾"
⅝"
5 ⅝"
NEW FACTS. Use of centre line
MODEL No. 5.
(Sand paper-block)
H
2"
4"
V
⅝
¾"
NEW FACTS. Drawing of two views (two planes) top view and front view; horizontal intersectional line; point of tangency
MODEL No. 6.
(Ruler)
V
P
1"
17"
NEW FACTS. Necessity of front and side (or end) view; vertical intersection line; how to represent a break.
MODEL No. 7.
(Steps.)
V
P
3 ¼"
¾"
¾"
1"
4 ¾"
1 ¾"
⅞"
6 ¼"
2 ½"
NEW FACTS. Sketching; front and side (or end) view; measuring and dimensioning spaces.
Lines.
Outlines; solid:
Dimension lines; ⅛" dash:
Extension and Construction lines; ⅛" dash:
Lines for Invisible edges; 1/32" dash:
Centre and Section lines; dot and dash:
Arrowheads: 6 ½" ¾" ⅛" 1/16"
Numbers: 1 2 3 4 5 6 7 8 9 0